Luiz Gustavo Batista Ferreira

Quality of Medicinal Plants market in Londrina, Brazil

Luiz Gustavo Batista Ferreira

Quality of Medicinal Plants market in Londrina, Brazil

Noor Publishing

Cover image: www.ingimage.com

Publisher:
Noor Publishing
is a trademark of
Dodo Books Indian Ocean Ltd. and OmniScriptum S.R.L publishing group

120 High Road, East Finchley, London, N2 9ED, United Kingdom
Str. Armeneasca 28/1, office 1, Chisinau MD-2012, Republic of Moldova, Europe
Printed at: see last page
ISBN: 978-620-7-47918-4

About the Author

Luiz Gustavo Batista Ferreira

Studied Agronomy at State University of Londrina

Master in Conservation Agriculture

Ph.D. Student in Agronomy

CONTENTS

ABSTRACT

The use of medicinal plants is a common practice around the world since the Ancient Egypt, when they used as food and medicine. Its use is of great importance especially in low-income populations without access to health resources. The purpose of this study was to evaluate the quality medicinal plant species traded in Londrina, Paraná state, Brazil. Medicinal plants were chosen on the basis of secretory structures or storage facilities of the active ingredients such as leaves, bark, flowers and fruits. Plant species chosen to perform the analyzes were phytosanitary quality: chamomile (*Chamomilla recutita* L. Rauschert), peppermint (*Mentha piperita*), anise (*Pimpinella anisum* L), hibiscus (*Hibiscus sabdariffa* L.), cinnamon (*Cinnamomum zeylanicum* Blume) and lemon balm (*Melissa officinalis* L.). The plant materials were purchased by five points or shops in city. The species purchased were transported to the laboratory, located at Londrina State University (UEL). Then, they were placed in sterile petri plates with PDA culture medium (potato, dextrose and agar) and 40 g of each plant species divided into four distinct points of the plate, and maintained at a constant temperature of 25°C for seven days. There were two replicates for species and point of sale. Were identified microorganisms harmful to human health as *Aspergillus flavus*, *Aspergillus niger*, *Aspergillus ochraceus*, *Fusarium* sp, *Nigrospora* sp, *Penicillium* sp, and *Rhizopus* sp. The six medicinal species acquired in trade from Londrina lower microbiological quality of medicinal plant. There is a high incidence of fungi harmful to human health as *Aspergillus flavus* and *Penicillium* sp in the six medicinal plants collected within five points of marketing.

Key words: Medicinal plants. Fungi. Contamination.

1. INTRODUCTION

Medicinal plants are plants that have biological activity with one or more active ingredients useful to health, with pharmacological action beneficial to the human organism and, for this reason, they are used as herbal medicines, especially in emerging regions and countries (FERREIRA, 1998; MATOS et al., 2018).

Studies indicate the relevance of the use of herbal medicines, to the detriment of the use of chemical medicines, either for social-economic reasons or for fear of synthetic medicines (CORDEIRO; FÉLIX, 2014; NETO et al., 2014; SILVA et al., 2015; VIEIRA; SOUZA; LEMOS, 2015).

Brazil has the largest biological diversity of medicinal plants in the world, arousing the interests of international scientific communities for the study, conservation and rational use of these resources (LORENZI, 2008; CAETANO et al., 2015).

For Nunes et al. (2003), the trade and use of medicinal plants are well known in Brazil. The market serves the consumer of herbal medicines in different ways, including the commercialization made in companies and local markets. However, Barbosa et al. (2010) point out that the lack of good practices in the production and marketing of medicinal plants used as phytotherapics, associated with the absence of vigilance, may compromise the microbiological quality.

Oliveira and Germano (1997) and Matos (2000) suggest that contamination of medicinal plants by fungal microorganisms may lead to alteration and / or destruction of the active principles of herbal medicines, compromising their health safety.

The contamination of phytotherapics occurs through the action of mycotoxins, which are produced by various species of fungi of genus Aspergillus spp, and can cause acute or chronic poisoning, with hepatotoxic, mutagenic, teratogenic and carcinogenic properties (CARVALHO; COSTA; CARNELOSSI, 2010).

Mycotoxins are secondary metabolites produced by fungi, at a biological stage of maturity. It is estimated that there are around 400 different mycotoxin types and are resistant to high temperatures (100°C to 130°C). Thus, even if the fungus is removed from a sample, mycotoxins will continue to act, which ultimately compromises a contaminated sample (BETINA, 1984).

The key fungal microorganisms synthesizing mycotoxins are: *Aspergillus fumigatus, Aspergillus flavus, Aspergillus niger, Aspergillus nudulans, Aspergillus terrus* and *Aspergillus ochaceus* (BETINA, 1984).

The purpose of this study was to evaluate the microbiological quality of medicinal plants used as herbal medicines marketed in Londrina, South of Brazil: Chamomile (*Chamomilla recutita* L.Rauschert), melissa (*Melissa officinalis* L.), sweet grass (*Pimpinella anisum* L.), cinnamon (*Cinnamomum zeylanicum* Blume), mint (*Mentha piperita*) and hibiscus (*Hibiscus sabdariffa*).

2. REVIEW ABOUT MEDICINAL PLANTS

The medicinal plant constitutes a therapeutic unit in which all the active ingredients form a phytocomplex that interacts with each other and with other apparently inactive molecules, this represents the integral pharmacological unit of the medicinal plant (JORGE, 2009). For Duarte et al. (2004), the biological activity of medicinal plants has been the subject of intense scientific investigation. Superior and aromatic plants are widely used in medicine, since they have a broad spectrum of activity and proven inhibition against bacteria and fungi, which makes medicinal vegetables in the formulations of medicines sold by the pharmaceutical industry (MATOS; LORENZI, 2002). Most of these properties are conferred by products of secondary metabolism, such as phenolic compounds and are present on a large scale in nature (SAAD et al, 2005).

Knowledge about medicinal plants often symbolizes the only therapeutic resource for many communities and ethnic groups. They have been commercialized since ancient times and arrived in Brazil along with Portuguese colonization (FIGUEIREDO, 2006). Even today, in the poorest regions of the country and even in large Brazilian cities, medicinal, aromatic and condiment plants are commercialized in open markets, popular markets and found in residential backyards (MACIEL et al., 2002).

Martins (2003) highlights important data from WHO (World Health Organization): 80 % of the world population has already used tea, while in Brazil, 90 %. The use of these plants has been stimulated, in part, by the growing demand of the industry for new natural sources of medicines and, on the other hand, due to the side effects caused by the synthetic drugs that stimulate the use of medicines of vegetal origin or, in many cases, because they represent the only source of medicines, especially in the most isolated and distant places (BERG, 1993), and as a response to immediate health problems (DEFILLIPS, 2001).

For Brito et al. (2009), the use of medicinal plants in curing diseases is demonstrated by the several examples cited in the literature regarding the use and efficacy of drugs of plant origin, such as studies using plants such as thyme, salvia, oregano and coriander (PRUTHI, 1980), cloves (HITOKOTO et al., 1980; MABROUK; El-SHAYEB, 1980), cinnamon, rosemary, rosemary (DAVIDSON, 1997), cumin, pepper, caraway (BEUCHAT; GOLDEN, 1989), garlic, onion (ABDOU et al., 1972; GANDHI; GHODEKAR, 1988), basil, turmeric, marjoram and anise (HITOKOTO et al., 1980) in the inhibition of human diseases.

Ferreira (1998) classifies medicinal plants in general terms as vegetables that have biological activity with one or more active ingredients useful for health. They are vegetables that have a pharmacological action beneficial to the human organism (MARTINS et al., 1995). They are of direct importance in phytotherapy as well as being a raw material for the pharmaceutical industry (KARL et al., 1997). The use of medicines, supplements and teas based on these plants is characterized as phytotherapy which, according to Sanguinetti (1989), there are numerous advantages of it, in its rational and sensible use, to treat diseases of the human body.

Classical phytotherapy consists of seeking to cure diseases through the use of medicinal plants. Several times, it is not an option for the population to use, but the only resource (CARRICONDE, 2002). This practice has been widely disseminated from generation to generation, in the form of teas, infusions and lickers (BRITO et al., 2009). The use of medicinal plants in Brazil, since before the beginning of Portuguese colonization, was used by native healers (MING et al., 2009). With regard to the empirical knowledge of popular herbal medicine, Morgan (1989) in his work, stated that such practice increases the scientific method through trial and error methods.

The WHO, on the other hand, refers to the medicinal plant as being "*any and all plants that contain, in one or more organs, substances that can be used for therapeutic purposes or that are precursors of semi-synthetic drugs*". The difference between medicinal plant and herbal medicine lies in the preparation of the plant for a specific formulation, which characterizes an herbal medicine (VEIGA JUNIOR et al., 2005). According to Duarte (2006), in the past, herbal medicine was more adopted by the poor in rural or urban areas, due to its easy availability and lower costs. Also according to the author, nowadays, the use of plants as a source of medicines is prevalent in developing countries such as Brazil, being an alternative solution to health problems (DUARTE, 2006).

According to Martins et al. (2003), the use of plants in Brazil for the treatment of diseases fundamentally presents influences from indigenous, African and European culture. The Indians used phytotherapy within a mystical vision. European influence, in turn, began with the arrival of the first priests of the Society of Jesus to Brazil. Traditional knowledge about ecology and the management of medicinal plants is fundamental for the rational and non-predatory use of natural resources (CASTELLANI, 2003).

Brazil has the greatest biological diversity in the world, with a rich flora, arousing the interests of international scientific communities for the study, conservation and rational use of these resources. Brazil, according to Plotkin (1991), is the country with the largest number of spermatophyte species in the world. Brazilian biomes contain one of the greatest diversities in the world (MENDONÇA et al., 1998), many of them with food and mainly medicinal value (ALMEIDA et al., 1998). For Duarte (2006), several Brazilian biomes have been used as natural drugs.

In the medical sector, tropical plants provide material for the production of analgesics, tranquilizers, diuretics, laxatives and antibiotics, among others. The worldwide commercialization of medicinal products of plant origin has a tendency to increase (MEYERS, 1983; PRINCIPE, 1985), since there are treatments and synthetic medicines that are too costly for the user.

The science of ethnobotany is considered a new scientific aspect (HAMILTON et al., 2003) and considered by Ford (1978) a scientific study of the direct interaction between the human species and plants, in dynamic systems (ALCORN, 1995) and has a wide interdisciplinary character scientific (ALBUQUERQUE, 2009). For Schultes and Reis (1995), ethnobotanical knowledge has existed since the origin of the species Homo sapiens.

The term ethnobotany was first used in 1895 by Harshberger, whom although he did not define it, pointed out ways in which it could be useful for scientific investigation. Since then, several definitions can be found for ethnobotany. Among the most recent ones, the following stand out: "*discipline that deals with the study and concepts developed by any society regarding the plant world*" - encompasses the way in which a social group classifies plants and the utility they give them; "*True scientific botany focused on the habitat and use of a specific ethnicity*" - carried out by someone trained in scientific botany, who would make correspondences between the western and local scientific classification; "*Botanical science that has a specific ethnicity*" - sees the culture of a society as everything that someone has to know or believe, in order to operate in a way acceptable to its members (MACIEL et al., 2002).

The ethnobotanical study consists of the evaluation of human interaction with all aspects of the environment, through surveys in traditional communities on the use of plants in the home pharmacopoeia and in the domestic economy. One of the pioneers in ethnobotanical studies, Richard Evans Schultes, dedicated to researching the flora of tropical America, reports on countless Brazilian plant species used by caboclos and indigenous people in the Amazon (MARTIN, 1995).

Several ethnobotanical studies have been developed in Brazil and in the world, seeking to know the popular medicine of traditional and / or contemporary peoples and the ways of organizing this knowledge, also looking for plants that effectively present a therapeutic activity and that consequently enables the discovery of new drugs (COUTINHO et al., 2002). Also, such studies allow inferring about the efficacy of the products that reach the market of production of teas, syrups, creams and others (ALEXIADES; SHELDON, 1996). The study of the uses of medicinal plants must take into account the social and cultural context in which these uses are inserted (HERRICK, 1983; ELISABETSKY, 1986 ETKIN, 1990; MACIEL et al., 2002).

Many plants frequently used by local populations have not yet been studied or their active ingredients have not yet been identified to validate them as medicines or to use them economically (BERG, 1993). Even so, many plants are used and commercialized today and Brazil, one of the countries with the greatest biodiversity in the world, reveals itself as an important and potential provider of a resource as valuable as medicinal plants. Examples of valuable plants in Brazil include the Indian Curare (*Digitalis purpurea*) used in the preparation of tea against dropsy, caused by heart failure before the action o on the heart muscle; *Drimys brasiliensis* was discovered, with stomatal properties; *Cinchona calisaya* used to cure malaria; *Cephaelis ipecacuanha* used to treat diarrhea, amoebic dysentery, chronic catarrh, hemorrhage and asthma, and *Carpotroche brasiliensis* with scientifically proven anti-inflammatory effects and whose oil extracted from the seed is used in the treatment of leprosy (COUTINHO et al., 2002).

The consumption of herbal medicines has increased considerably in the last two decades, both in developed and developing countries. In Europe alone, the herbal medicine market reaches about 7 billion USD a year, with Germany accounting for 50 % of this value (ABIFISA, 2012). But in Brazil, even given the importance of this market, there are no official data on how much the herbal medicine industry moves.

It is estimated at around 1 billion Brazilian Reais / year (ABIFISA, 2012) with a tendency to grow more and more. Due to the existence of problems related to the quality and regularity of supply, most pharmaceutical companies, 70 % transnational companies based in Brazil, prefer to import their raw materials. Therefore, although there is a demand for medicinal plants and potential for service, there is unpreparedness on the part of farmers / collectors to meet the demands of the market and the production chain in general.

According to Nunes et al. (2003), the trade and use of medicinal plants are well known and discussed in Brazil and worldwide. The market serves the consumer of this type of product in different ways, including sales made in companies, markets and herbalists (BRANDÃO; FREIRE; VIANNA-SOARES, 1998a). In different countries, folk medicine using plants is widely practiced by healers and small herbalists, and is expanding rapidly. Technically prepared products from medicinal plants can also be mentioned, which are herbal medicines, a market that has also expanded (CALIXTO, 2000; SUZUKI, 2002).

The WHO has described the profile of this user, revealing that the majority of the world population, living mainly in developing countries, does not have access to modern medicine and, consequently, to synthetic and herbal medicines, then resorting to folk medicine (FARNSWORTH et al., 1985) which is available, among other forms, as products sold by farmers and herbalists or extracted from sources close to places of residence. Several socioeconomic factors have been pointed out to explain this phenomenon, among which they range from the high cost of medicines to the fads (BRANDÃO; FREIRE; VIANNA-SOARES, 1998b).

The products marketed should respect the criteria of adequacy to health and meet the binomial safety and effectiveness, as is required of other products. What has been observed, however, is a poor quality of the commercialized medicinal plants, the lack of pharmacovigilance (BRANDÃO; FREIRE; VIANNA-SOARES,

1998b) and the use of species not fully evaluated from the pharmacological and toxicological point of view.

In Brazil, this occurs mainly with native plants (RITTER et al., 2002), which are the most commercialized because they are obtained in an extractive manner. According to Barbosa et al. (2010), the lack of good practices in the production and commercialization of medicinal plants and herbal medicines, associated with the absence of pharmacovigilance, contributes to significant drops in the quality of herbal medicines. To maintain the quality of a phytotherapic, adequate identification, cultivation, collection, handling, drying, storage and transport practices are necessary, as indicated in Resolution RDC nº 306, of December 7, 2004, by ANVISA.

Medicinal, aromatic and condiment plants, as well as other plants of economic interest, can be affected by phytopathogens, such as rust, leaf spots, and powdery mildew (RUSSOMANO; KRUPPA, 2010; LIMA et al., 2004). Such diseases decrease the quality of medicinal plants and the yield of crops in general (CUMMINS; HIRATSUKA, 2003; HAWKSWORTH et al., 1995; AGRIOS, 1997; URBEN et al., 1987; SOUZA et al., 1995), and the main climatic factors that contribute to the incidence of plant diseases are temperature and high relative humidity in general (MELCHING et al., 1989; ALVES et al., 2006; TSCHANZ, 1984; DEL PONTE et al., 2006).

For the authors Russomano and Krupa (2010) control of fungal pathologies of the aerial part of medicinal, seasoning and aromatic plants should be done, preferably, by the use of resistant varieties and by the use of inoculum-free seeds, seedlings and propagating material, in places without occurrence of diseases. The use of healthy and highly vigorous seeds avoids the introduction of pathogens in the beds, nurseries and in the field and, when sown at an adequate depth, allow the rapid emergence and growth of the plant, remaining for a shorter time susceptible to causal agents.

When there are no resistant varieties, the use of cultural control techniques can help reduce the damage caused by diseases. In this context, the use of pesticides (fungicides) should be avoided, due to the impact they cause on the environment and human health, by the pollution of their chemical residues. In addition, there is insufficient information on the effect of the fungicide on the metabolism of the active, therapeutic and flavoring principles, even in the presence of toxic residues in the extracts used for medicinal and culinary purposes.

Nowadays, several and innovative strategies have been used to control diseases in agriculture, such as biological control using Trichoderma, use of plant extracts, elimination of cultural remains, crop rotation (with non-host cultures of pathogens), fertilization balanced, control of invasive species, appropriate and recommended spacing, cultivation of plants in places that present soil with good drainage and pruning of branches of trees and shrubs (RUSSOMANO; KRUPPA, 2010).

However, there are still no scientific studies or records to prove that these diseases affect the human organism. Medicinal, aromatic and condiment plants can host harmful pathogens such as fungi of the *Aspergillus*, of the species *Fumigatus nige*r (filamentous fungus), *Penicillium* and *Fusarium*, and such microorganisms can be marketed together with medicinal plants and the final consumer, through teas, for example, can ingest these fungi causing health complications through such microorganisms (TORTORA et al., 2000).

The increase in tea consumption is worrying considering that the inspection of trade in such products is precarious; thus representing several risks to the consumer, the main ones being the wrong identification of the plant species, erroneous indications of use, use of poor quality plants and toxic species (ESPÍNOLA, 1997).

According to Oliveira et al. (1997), Correa Junior et al. (1994) and Matos (2000), the contamination of medicinal species by fungi can lead to alteration and / or destruction of the active principles, thus causing loss of safety and efficiency in use, in addition to of representing a risk due to the production of toxic substances (mycotoxins), thus making them unfit for consumption, regardless of the level of contamination.

Aflatoxins, secondary metabolites produced by several species of fungi of the *Aspergillus* genus, represent mycotoxins of great medical and sanitary interest due to the high occurrence and high toxicity, causing acute or chronic intoxications, with hepatoxic, mutagenic, teratogenic and cancerous properties (ROITMAN et al., 1988) and are thermostable with a melting point around 269°C (CARVALHO et al., 2010). They are classified, according to the International Agency for Research on Cancer (IARC), as a carcinogenic agent for humans. For Rosa (2012), there are morphological evidences in the fungi that characterize the production of aflatoxins, such as the presence of blastoconides, chlamydocides and sclerocytes.

Aflatoxin B1 is one of the aflatoxins produced by the fungus Aspergillus flavus, which grows mainly in food. This fungus can develop in food, having a very high incidence in the storage, mainly of cereals, peanuts and manioc. Its development is rapid, reaching very high levels in favorable environments with ideal temperature and humidity (BIS, 2012). Caldas et al. (2002) states that it is not uncommon for people to be intoxicated by mycotoxins in food and medicinal species in the form of tea.

The use of plants for therapeutic purposes without guarantee of safety and efficacy is aggravated by indiscriminate and uncontrolled trade, representing danger to the consumer, the risk of purchasing and using poor quality products, in addition to inappropriate post-harvest storage conditions (ARAÚJO; OHARA, 2000).

The *Rhizopus* genus is also considered pathogenic because it is an opportunistic saprophyte (ESTEVES et al., 1990; BUGNO et al., 2005). *Aspergilluses* are mycoses produced by opportunistic fungi of the Aspergillus genus, mainly A. *fumigatus* and less frequently by A. *flavus*, A. *nidulans*, A. *niger*, A. *terrus* and A. *ochraceus* (MEZZARI; CAUDURO, 1996). In addition, molds caused by *Fusarium* sp., *Aspergillus* sp. and *Penicillium* sp are also part of this range of contaminating agents, and are also producers of mycotoxins (ZARONI et al., 2004), representing a risk to the user of medicinal plants in the form of tea.

Zygomycosis is an occasional disease, with cutaneous, subcutaneous and systemic clinical manifestations that is caused by opportunistic fungi that become pathogenic, being the species *Rhizopus oryzae*, R. *rhizoposiformis* and R. *microsporus* pathogenic agents of this disease (MEZZARI; CAUDURO, 1996). Microorganisms can contaminate vegetables during the pre- and post-harvest stages. In pre-harvest, the main sources of contamination are soil, fertilizers not properly composted, contaminated irrigation water, water used for the application of fungicides and insecticides, dust, insects, domestic and wild animals and human manipulation (ZARONI et al., 2004).

Sources of post-harvest contamination include human manipulation, inadequate or insufficient cleaning of harvesting equipment, dirty and contaminated transport packaging, animals, insects, dust, contaminated washing water, transport vehicles and contamination of equipment used in post-processing. harvesting, drying and storage, performed inappropriately, without any type of control (BRACKETT, 1999; BEUCHAT, 2002; KALKASLIEF-SOUZA et al., 2009; SATOMI et al., 2005).

Matos (1998) strongly emphasizes and emphasizes the essential care that must be taken in handling plant materials, such as correct drying, maintaining minimum hygiene conditions to avoid the development of molds and to prevent the material from coming into contact with the ground, dust or even rain, which favors the development of microorganisms in medicinal plants.

Silva et al. (1995) warns about the risk that consumers run due to the possibility of the final product presenting fungal contamination in the most different commercial points (BUGNO et al., 2006).

Thus, it is necessary to establish adequate measures of hygienic-sanitary control to guarantee the quality and safety of this type of product, from collection, storage, handling to the final product (homemade medicines or herbal medicines). In addition to these adjustments, the microbiological control of plant material and its derivatives must be more rigorous and efficient on the part of the manufacturers and more required by the responsible government agencies to guarantee the safety, efficacy and quality of the products, being, therefore, extremely necessary, the consumer follows legal recipes and that the products contain information about their quality, avoiding a likely contamination of users of medicinal and aromatic plants (FARMACOPÉIA BRASILEIRA, 2005; OLIVEIRA et al., 1997; DINIZ et al., 1997; MATOS , 2000).

Chamomile (*Chamomilla recutita* L.Rauschert), from the Asteraceae family, is native to Europe and acclimated in some regions of Asia and countries in the American continent, with the largest area of cultivation. The state of Paraná stands out as the largest producer, with a cultivated area of 700 ha and a production of around 260 t, which serves 74 % of national consumption, which is 350 t (DALLA COSTA, 2001; CORREA JÚNIOR et al., 2003). Part of the chamomile consumed in the country is imported from Argentina, the main world producer (CURIONI, 2004).

Chamomilla recutita is herbaceous, annual, aromatic with flowers grouped in chapters, being widely cultivated practically everywhere in the world (LORENZI, 2008). According to Lorenzi (2008), it is considered one of the oldest-used plants by traditional European medicine, now included as an official in the Pharmacopoeias of almost all countries. It is used in both scientific and popular medicine, in the form of infusion and decoction (cooking of the chapters), to facilitate the elimination of gases, fight colic and stimulate appetite, also acting topically by applying compresses of the infusion, still hot on the abdomen in the treatment of colic in children.

Melissa oficinalis L, from the Verbanaceae family, in the region of Londrina popularly known as lemongrass, is a sub-shrub with variable morphology, with an average height of 150 cm. Its branches are thin, whitish, arched, long and brittle, the leaves are whole, opposite, with serrated edges and acute apex, of 3-6 cm in length, the flowers are blue-colored, gathered in capituliform axillary inflorescences of short axis and variable size (LORENZI, 2008). For Ming (1992), this species is cultivated and used throughout Brazil for its analgesic, antispasmodic, soothing, sedative and cytostatic properties and its therapeutic effects have already been scientifically proven. According to Scheffer (1992), little information is available regarding the agronomic aspect, with the need for scientific studies that reveal the behavior of medicinal species when subjected to production techniques without affecting the therapeutic value of the plant considering the fact that the active principles may present alterations according to the cultivation techniques (MADUEÑO BOX, 1973).

For Lorenzi (2008), *Pimpinella anisum* L., belonging to the Apiaceae family, known for fennel, is native to Asia and cultivated in southern Brazil; it is considered aromatic, annual, erect, up to 50 cm in height, it has leaves composed of various shapes, cracked and white flowers, arranged in umbels. The fruits are achenes, with a sweet taste and strong smell, in addition to being dry (mericarps) being used since ancient times, as a stimulant of digestive functions, to eliminate gas, fight colic, headache and stimulate lactation. The use of tea is internationally approved as a simple measure against the cold, cough and bronchitis, fever, colic, inflammation in the mouth and throat, poor digestion and loss of appetite (LORENZI, 2008).

The medicinal plant *Cinnamomum zeylanicum* Blume (cinnamon) belongs to the Lauraceae family, and its center of origin is Sri Lanka and in southeastern India. It is an important spice and used in the perfumery and beverage industries (JAYAPRAKASHA et al., 2003), grown in different regions around the world. It is also used in the pharmaceutical industry, since it confers health benefits due to its active ingredients (KHAN et al., 2003). According to Shan et al. (2005), chemical analyzes prove that this plant is rich in polyphenols and volatile oils.

Lorenzi (2008) describes the cinnamon plant as an aromatic tree from 6 to 12 meters high, with opposite, ovate and trinervated leaves. The flowers are numerous and gathered in branched racemes and arranged in terminal panicles, greenish-yellowish in color. Boorhem (1999) recommends the popular use for treating childhood diarrhea, flu, worms, toothache and bad breath. Pharmacological tests have shown that cinnamon essential oil has antibacterial activity against microorganisms that cause diseases in the human organism's respiratory system (LORENZI, 2008).

The hibiscus (*Hibiscus sabdariffa* L.) is an annual or biennial, upright, branched, bushy stem, 80-140 cm tall and native to the African continent, with alternating, long-petiolate leaves, whole at the base of the plant and 3 or 4 located at the apex, with toothed margins, 5-12 cm long; the flowers are solitary, axillary, yellow in color, and the fruits are capsules coated by hybrids (LORENZI, 2008). For Panizza (1997), hibiscus can be used in a therapeutic way, as a diuretic, laxative, stoma, calming and anti-scorbutic with the use of chalices in teas, soft drinks, jams and condiments.

Mint (*Mentha piperita*) is used for medicinal and food purposes (LORENZI; MATOS, 2002). Its cultivation has economic importance, due to its capacity to produce and store essential oil, whose main constituent is menthol, used in oral hygiene products, pharmaceuticals, cosmetics and food. Menthol also has high antifungal activity, in addition to antibacterial potential (SOUZA et al., 1991).

3. MATERIAL AND METHODS

The experiment was carried out at the phytopathology laboratory of UEL - Londrina State University, South of Brazil, at coordinates 23°23' South Latitude and 51°11' West Longitude Greenwich, mean altitude 565 m and Cfa type according to the Köppen climate classification (IAPAR, 2000).

The medicinal species used as phytotherapics were chosen based on the plant organs of the plants that present secretory or storing structures of the active ingredients such as leaves, bark, flowers and fruits. We also took into consideration the best known and commercialized plants in Londrina, Brazil.

The species chosen to perform the analyzes were: chamomile (*Chamomilla recutita* L. Rauschert), mint (*Mentha piperita*), fennel (*Pimpinella anisum* L), hibiscus (*Hibiscus sabdariffa* L.), cinnamon (*Cinnamomum zeylanicum* Blume) and melissa (*Melissa officinalis* L.).

Plant materials were purchased from five commercial locations in Londrina, PR: For reasons of ethics, locations do not have names, only "location 1", "location 2", "location 4", "location 4" and "location 5". Two samples of each medicinal species were acquired at each site, 12 samples per site and 60 samples in total.

After acquisition, the samples were transported to the phytopathology laboratory of Londrina State University.

The criteria for choosing the locations were: most populous region of the municipality, in the northern zone (site 1, open market); central stores (local 2, local 3 and local 4) and a local specializing also in the central region in natural remedies (homeopathic pharmacy, local5). Were specific locations of marketing of herbal medicines in Londrina, Brazil. In the results, locations are shown with the letter "L" and the location number from 1 to 5.

The petri dishes used in the work were washed with 70 % alcohol and sterilized at a temperature of 170°C for a period of 60 minutes, according to the methodology adopted by Mendonça; Hidalgo; Chaves (2009).

Figure 1. Samples of herbal medicines acquired at points of commerce, in Londrina, Brazil.

Source: Author (2019)

Figure 2. Petri dishes, in the phytopathology laboratory of Londrina State University, separated and washed to perform the analyzes.

Source: Author (2019)

After sterilization, the Petri dishes were taken to the flow chamber for insertion of potato; dextrose and agar and the addition of 40 g of each sample divided into four distinct points of the plate. Then, the plates were sealed and left at a constant temperature of 25°C for seven days in the absence of light (BRASIL, 2003).

After seven days, Petri dishes were evaluated. Lactophenol blue staining slides were prepared for fungal tincture and observed under the laboratory optical microscope.

The identification of the fungi was performed according to identification keys proposed by Barnet and Hunter (1972), considering the structural characteristics of the microorganisms.

According to the presence or absence of microorganisms, the samples were classified into "contaminated samples, X-marked in spreadsheets" and "uncontaminated samples".

3. RESULTS AND DISCUSSION

Was verified microorganisms in the samples: *Alternaria sp., Aspergillus flavus, Aspergillus niger, Aspergillus ochraceus, Fusarium sp., Penicillium sp., Rhizopus sp.*

In the scientific literature, these microorganisms are considered potentially toxic because they are mycotoxin synthesizers (EMBRAPA, 2007; INANAKA et al., 2010; FERREIRA; BARROS; FORTUNA, 2017).

The percentage, observed in Table 1, was established based on the total number of microorganisms found in the samples. In all, the presence of seven fungi was verified, therefore, this value represents 100%. From this, the percentages of microbiological contamination were established for each site and for each medicinal species.

It is also noteworthy that the "X" marks the presence of pathogen in both samples, referring to the site in question, and the absence of "X" implies the absence of microorganisms in the two samples acquired from that site.

The results were verified in this work are presented in Table 1.

The main aspect observed in Table 1 was that the five sample collection points presented fungal microbiological contamination. Souza and Maciel (2010) suggest minimal contamination tolerance in a colony forming unit in herbal medicines. However, the present work points out that only the presence of the microorganism already indicates microbiological contamination of the herbal medicine, with potential risk to human health.

Table 1: Presence of toxigenic microorganisms verified in medicinal herbs samples acquired in Londrina, Brazil.

	Microorganisms	Local (medicinal herbs marketing)						Microorganisms	Local (medicinal herbs marketing)				
		L1	L2	L3	L4	L5			L1	L2	L3	L4	L5
Pimpinella anisum L.	*Alternaria sp*	X		X	X		*Chamomilla recutita* L. Rauschert	*Alternaria sp*	X		X		
	Aspergillus flavus							*Aspergillus flavus*	X	X	X	X	X
	Aspergillus niger	X				X		*Aspergillus niger*	X	X		X	
	Aspergillus ochraceus	X		X				*Aspergillus ochraceus*		X			
	Fusarium sp	X			X	X		*Fusarium sp*	X		X		
	Penicillium sp	X	X			X		*Penicillium sp*	X	X			
	Rhizopus sp	X	X		X	X		*Rhizopus sp*	X			X	
	% local*	85,8	28,6	28,6	42,9	57.2		% local*	85,8	57.2	42,9	42,9	14,3
Melissa officinalis L.	*Alternaria sp*						*Hibiscus sabdariffa* L.	*Alternaria sp*					
	Aspergillus flavus	X						*Aspergillus flavus*					X
	Aspergillus niger	X		X	X			*Aspergillus niger*	X	X		X	
	Aspergillus ochraceus			X				*Aspergillus ochraceus*					
	Fusarium sp	X	X	X	X			*Fusarium sp*	X	X			
	Penicillium sp	X						*Penicillium sp*	X			X	X
	Rhizopus sp		X			X		*Rhizopus sp*			X	X	
	% local*	57.2	28,6	42,9	28,6	14,3		% local*	42,9	28,6	14,3	42,9	28,6
Cinnamomum zeylanicum	*Alternaria sp*	X					*Mentha piperita*	*Alternaria sp*			X		X
	Aspergillus flavus		X					*Aspergillus flavus*		X			
	Aspergillus niger	X	X			X		*Aspergillus niger*	X	X			X
	Aspergillus ochraceus		X					*Aspergillus ochraceus*	X		X		
	Fusarium sp	X			X			*Fusarium sp*		X			X
	Penicillium sp	X			X			*Penicillium sp*	X		X		X
	Rhizopus sp			X				*Rhizopus sp*	X	X		X	X
	% local*	57.2	42,9	12,5	28,6	14,3		% local*	57.2	57.2	42,9	14,3	71,5

* Percentage of samples contaminated by pathogenic microorganisms at the locations of marketing of herbal medicines (L1, L2, L3, L4, L5).

The highest percentage of contamination was found at site 1, where it ranged from 42.9 % in hibiscus to 85.8 % for sweet grass and chamomile. The lowest percentage of contamination, in turn, was at site 4, which ranged from 12.5 % to 42.9 % of contaminated samples.

Aspergillus flavus contamination was observed in the samples of commercialized chamomile flowers, a pathogen capable of producing carcinogenic compounds called aflatoxins.

Figure 3. Presence of *Aspergillus flavus* in chamomile (*Chamomilla recutita* L.Rauschert) samples.

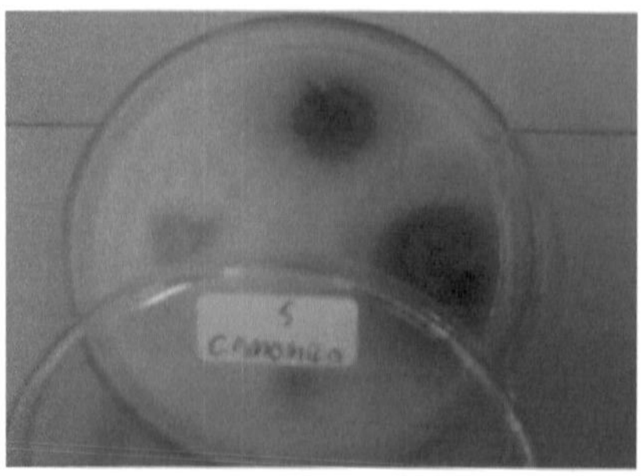

Source: author (2019)

In was verified, *Alternaria sp.* in 3 locations (Figure 4). The presence of *Alternaria sp.* on all shows of mint. There are no studies indicating that this genus produces mycotoxins.

Figure 4. Presence of *Alternaria sp.* in samples of fennel (*Pimpinella anisum*)

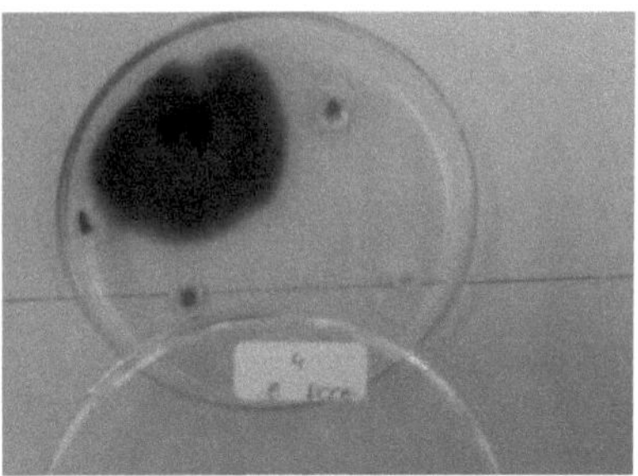

Source: author (2019)

The medicinal species mint presented the most contaminated samples. There were 38 samples containing the 7 toxigenic microorganisms found in the work: *Alternaria sp., Aspergillus flavus, Aspergillus niger, Aspergillus ochraceus, Fusarium sp., Penicillium sp., Rhizopus sp.*

Figure 5. Presence of *Aspergillus niger, Rhizopus sp. Mintha piperita samples.*

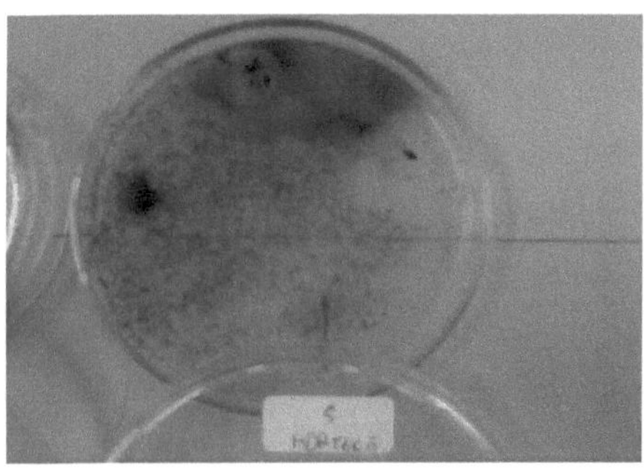

Source: author (2019)

The frequency of harmful pathogens identified in hibiscus samples is high. Of all samples, only the fungus *Aspergillus ochraceus* was not found and identified in them. This implies reduced reliability and health safety for the use of this medicinal species as an herbal medicine.

The hibiscus samples presented 24 contaminated samples, from which 14 samples were identified *Penicillium sp.*, As shown in Figure 6. *Alternaria sp.* and *Aspergillus ochraceus* were not found in any sample.

Figure 6. Samples of hibiscus (*Hibiscus sabdariffa* L.) with presence of *Penicillium sp.*

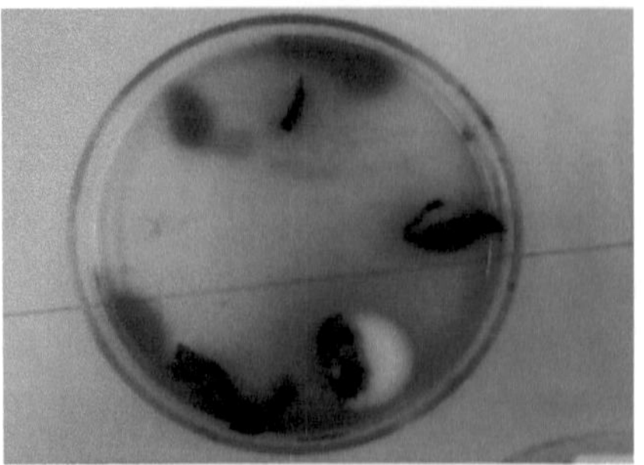

Source: author (2019)

The *Aspergillus niger* fungus was found in all 60 samples for all medicinal plants under study, especially the samples of melissa (*Melissa officalis* L.) and cinnamon bark (*Cinnamomum zeylanicum* Blume) leaves, whereas *Aspergillus flavus* was observed in all chamomile samples. Both microorganisms are mycotoxin producers, compromising the use of this herbal medicine in terms of microbiological quality due to the high aflatoxin-producing mycoflora found in chamomile samples (FERREIRA; BASTOS; FORTUNA, 2017).

Figure 7: *Aspergillus niger* identified in the leaves of *Melissa officinalis* L.

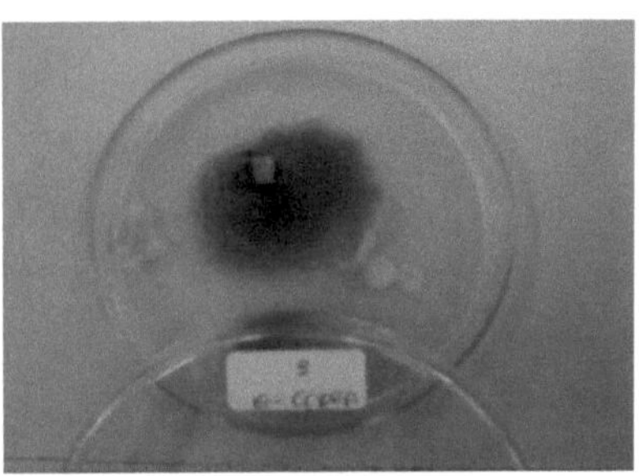

Source: author (2019)

Fischer and Saito (1988), Fischer, Ohara and Saito (1993), Amaral (1999) and Araújo and Ohara (2000) observed samples contaminated by *Aspergillus* spp. and *Penicillium sp*, which is in agreement with the results observed in Table 1.

Bennett; Klich, 2003, Leung et al., 2006, Bhat; Miller, 1991 and Dawson, 1991 suggest that the presence of *Aspergillus* spp. and *Penicillium* sp in samples, indicate microbiological contamination.

The microorganisms found in the samples of this work Aspergillus spp. and *Penicillium* sp. can cause diseases in immunosuppressed, such as allergies, and pose risks to those who handle such herbal medicines, given the possible inhalation during the production process (MAXIMINO et al., 2011).

The authors Mendonça; Hidalgo; Chaves (2009) identified *Aspergillus niger*, *Aspergillus flavus*, *Penecillium* sp. In phytotherapic samples acquired from commercialization points in the municipality of Manaus, capital of Amazonas. and *Ryzopus sp*, which were also identified in this work.

The materials from site 1 (open market) do not show any hygiene care, with commercial presentation (packaging and labels, generic and visually poor). Similar situation was observed by Mendonça; Hidalgo; Chaves (2009), who point out that more than 60% of herbal medicines marketed in the municipality of Umuarama, in Northwestern of Paraná state, present microbiological contamination in free market places.

Dall'agnol and Alves (1998) found that medicinal species commercialized in the city of Curitiba, present high rates of microbiological contamination.

Souza and Maciel (2010) show that whole-leaf medicinal plants for consumption in the municipality of Campinas, SP, presented contamination of 92% of the samples by fungi of the genus *Fusarium sp., Penicillium sp.* and *Rhizopus* sp., in phytorapic samples, the same genera found in this work, which indicates that the results presented are in accordance with the literature.

The evaluation of the phytosanitary quality of medicinal plants in Brazil, in the works carried out by Fischer and Saito (1988), Fischer et al. (1993), Santos et al. (1995), Dallagnol and Alves (1998), Amaral (1999) and Araújo and Ohara (2000) demonstrated that the main microorganisms involved in the contamination of such products are fungi, with a high occurrence of the genera *Aspergillus* and *Penicillium*. For Bugno et al. (2006), the contamination of products of plant origin by fungi is somehow related to the inadequate conditions of harvest and post-harvest. This situation is aggravated by consumers of plants for medical use who buy them in markets and fairs, where the intense circulation of people, along with packaging with inadequate packaging, exposing the plant material to heat, humidity, dust and products of another nature, such as genera perishable foodstuffs, effectively contribute to the additional contamination of the material.

4. CONCLUSIONS

The microorganisms found in the phytotherapic samples were from potentially toxigenic fungi, especially *Aspergillus* spp, producers of mycotoxins that compromise the microbiological quality of medicinal species used as phytotherapeutic agents.

The percentage of contamination among commercial establishments varied, but the highest rates were registered for "local 1", free fair.

Phytotherapics are natural alternatives to treat diseases, but their use should be cautious, because of the risk of mycotoxin contamination. In this context, the rigor and legal safety regulations for natural remedies should be the same for chemical medicines.

Inspection and compliance with the normative instructions of the field for storage as well as sanitary inspection at the points of sale of phytotherapics, especially in open markets, where the products are exposed to the outdoors, are more likely to be contaminated by *Aspergillus* spores.

The results of this work were similar to other research in other regions of Brazil, indicating that microbiological contamination of herbal medicines is a national problem, not just local or regional, and show the need for phytotherapics trade control and control policies.

5. REFERENCES

ABDOU, I. A.; ABOU-ZEID, A. A.; El-SHERBEENY, M. R.; ABOU-EL-GHEAT, Z. H., Antimicrobiol activities of *Allium sativum, Allium cepa, Raphanus sativus, Capsicum frutescens, Eruca sativa, Allium kurrat* on bacteria. **Quality of Plant Material.**, p. 29-35, 1972.

ABIFISA - Associação Brasileira das Empresas do Setor Fitoterápico, **Suplemento Alimentar e de Promoção da Saúde**, 2012

AGRIOS, N.G. **Plant pathology**. 4. ed. San Diego: Academic Press, p.635., 1997.

ALBUQUERQUE, U.P. Quantitative ethnobotany or quantification in ethnobotany? **Ethnobotany Research and Applications**, 7: 1-2, 2009.

ALCORN, J.B. The scope and aims of ethnobotany in a developing world. In: R.E. Schultes and S.V. Reis (eds.). **Ethnobotany: evolution of a discipline**. Cambridge, Timber Press, p.23-39, 1995.

ALEXIADES, M.N.; SHELDON, J.W. **Ethnobotanical Research:** A Field Manual. Bronx, New York, The New York Botanical Garden, 1996.

ALMEIDA, S.P.; PROENÇA, C.E.B.; SANO, S.M.; RIBEIRO, J.F. **Cerrado:** Espécies vegetais úteis. Planaltina, EMBRAPA-CPAC, Distrito Federal, p.464, 1998.

ALVES S.A.M; FURTADO, G.Q; BERGAMIN FILHO, A. Influência das condições climáticas sobre a ferrugem da soja. In: Zambolim L (Ed.). **Ferrugem asiática da soja**. Viçosa MG. Suprema Gráfica e Editora Ltda. p. 37-59, 2006.

AMARAL, F. M. M. Frutos de *Luffa operculata* (L.) Cogn.: avaliação da comercialização e controle de qualidade de amostras adquiridas em mercados de São Luís/MA. **Revista Brasileira de Farmacognosia.** São Luis. Universidade Federal do Maranhão, p.27-31, 1999.

ARAÚJO, A. L. A.; OHARA, M. T. Qualidade microbiológica de drogas vegetais comercializadas em feiras de São Paulo e de infusos derivados. **Revista Brasileira de Ciências Farmacêuticas**, v.36, n.1, jan./jun., p.129- 136, 2000.

BARBOSA, C.K.R.; COSTA, J.P.R.; BONFIM, F.P.G.; ALMEIDA, A.C.; MARTINS, E.R. Qualidade microbiológica de plantas medicinais cultivadas e comercializadas em Montes Claros, **Instituto de Ciências Agrárias, Universidade Federal de Minas Gerais Biotemas**, Minas Gerais, p. 77- 81, 2010

BATALHA, M. O.; NANTES, J. F. D.; ALCÂNTARA, R. L. C.; MING, L. C.; CASTRO, D. M. de; LOURENZANI, A. E. B. S.; MACHADO, J. G. C. F.; RIBEIRO, P. M. T. Plantas medicinais no Estado de São Paulo: situação atual, perspectivas e entraves ao desenvolvimento. **Revista Brasileira de Plantas Medicinais**, São Paulo (SP), p. 64-69, 2007.

BERG, M.E. **Plantas medicinais na Amazônia:** contribuição ao seu conhecimento sistemático 2 ed. Rev. E aum. - Belém: Museu Paraense Emílio Goeldi ,1993.

BETINA, V. **Mycotoxins**: production, isolation, separation, and purification. Amsterdam: Elsevier, 1984. 528 p.

BEUCHAT, L. R. e GOLDEN, D. A., Antimicrobials ocorring naturally in foods. **Food Technol.**, U.S.A, p. 134-142, 1989.

BEUCHAT, L. R. Ecological factors influencing survival and growth of human pathogens on raw fruits and vegetables. **Microbes and infection**, Paris, v. 4, p. 413-423, 2002.

BIS. **Organização Logística Interna Otimização de Processos Utilidades Energia Racionalização da Produção Sistemas de Gestão da Qualidade e Ambiental.**

BOORHEM, R.L. **Segredos e Virtudes das Plantas Medicinais**, Reader Digest, Brasil, Rio de Janeiro, p.416, 1999.

BRACKETT, R.E. Incidence, contributing factors, and control of bacterial pathogens in produce. **Postharvest Biology and Technology**, Amsterdam, v. 15, p. 305-311, 1999.

BRANDÃO, M. G. L.; FREIRE, N.; VIANNA-SOARES, C. D. Vigilância de fitoterápicos em Minas Gerais. Verificação da qualidade de diferentes amostras comerciais de camomila. **Cadernos de Saúde Pública**, v.14, n.2, p.693-700, 1998 a.

BRANDÃO, M. G. L.; FREIRE, N.; VIANNA-SOARES, C. D. Vigilância de fitoterápicos em Minas Gerais. Verificação da qualidade de diferentes amostras comerciais de camomila. **Cadernos de Saúde Pública**, v. 14, n. 3, p. 613-616, 1998 b.

BRASIL. Ministério da Agricultura, Governo Federal. **Comunicado Técnico 101**, 1 edição, Brasília,Distrito Federal, 2003.

BRASIL. Ministério da Agricultura, Governo Federal. **Manual de Análise Sanitária de Sementes**, Brasília ,Distrito Federal, p.202, 2009

BRITO, V.F.S.; DANTAS, I.C.; DANTAS, G.D.S. Plantas medicinais utilizadas pela comissão de mulheres na zona rural no município de Lagoa Seca - PB, **Revista de Biologia e Farmácia**, vol3, n 1,Lagoa Seca, p.112-121, 2009.

BUGNO, A.; BUZZO, A.; A.; NAKAMURA, C. T. Avaliação da contaminação microbiana em drogas vegetais. **Revista Brasileira de Ciências Farmacêuticas**. Vol. 41, n. 4, p.491-497, 2005.

BUGNO, A.; ALMODOVAR,A.A.B.; PEREIRA, T.C.;PINTO, T.J.A.; SABINO,M **Currence of toxigenic fungi in herbal drugs,** Instituto Adolfo Lutz, São Paulo, SP, Brasil; Faculdade de Ciências Farmacêuticas, Universidade de São Paulo, São Paulo, SP, Brasil,p.47-51, 2006

CALDAS, E. P.; SILVA, S. C.; OLIVEIRA, J. N. Aflatoxinas e ocratoxinas em alimentos e riscos para a saúde humana. **Revista de Saúde Pública**, São Paulo, v. 36, n. 3, 2002.

CALIXTO, J. B. Efficacy, safety, quality control, marketing and regulatory guidelines for herbal medicines (phytotherapeutic agents). **Brazilian Journal of Medical and Biological Research,** v. 33, p. 179-189, 2000.

CARRICONDE, C. **Introdução ao uso de fitoterápicos nas patologias de APS.** Olinda: Centro Nordestino de Medicina Popular, 2002.

CARVALHO, L.M.; COSTA, J.A.M.; CARNELOSSI, M.A.G. **Qualidade em Plantas Medicinais**, Aracajú, p.56, dez. 2010.

CASTELLANI, D. C, **Plantas medicinais e aromáticas:** produtos florestais não Madeireiros (PFNM) In. COELHO, M. F. B.; JÚNIOR, P. C.; DOMBROSKI, J. L. D. (orgs.) **Diversos olhares em etnobiologia, etnoecologia e plantas medicinais**: anais do I Seminário Matogrossense e II Seminário Centro-Oeste de plantas medicinais- Cuiabá: Unicen, p.199-212, 2003.

CORREA JUNIOR, C.; MING, L. C.; SCHEFRER, M. C. **Cultivo de plantas medicinais e aromáticas.** 2ª ed. Jaboticabal: FUNEP, p 162, 1994.

CORREA JÚNIOR, C.; EMPINOTTI, A.L.; SCHEFFER, M.C.; GRAÇA, L.R. **Estudo da cadeia produtiva de plantas medicinais, condimentares e aromáticas.** Curitiba: EMATER-PR. p.26, 2003.

COUTINHO, D. F.; TRAVASSOS, L. M. A.; AMARAL, F. M. M. do. Estudo etnobotânico de plantas medicinais utilizadas em comunidades indígenas no estado do Maranhão – Brasil. **Visão Acadêmica**, Curitiba, v. 3, n. 1, p. 7-12, Jan.-Jun./2002

CUMMINS, G.B. e HIRATSUKA, Y. **Illustrated Genera of Rust Fungi**. 3 ed. rev. Saint Paul:American Phytopathological Society, p. 240, 2003.

CURIONI, A.O. Cultivation and post harvest process for aromatic and medicinal plants in Argentina. In: **International symposium breeding research on medicinal and aromatic plants, 3, latin American symposium on the production of medicinal and aromatic plants and condiments**, Campinas. Program and Abstracts, Campinas: UNICAMP/IAC/CEFET/ UNESP, p.5, 2004.

DALLA COSTA, M.A. Processo de produção agrícola da cultura da camomila no município de Mandirituba, PR. (**Dissertação de mestrado**), UFPR, Curitiba, p.69, 2001.

DALL'AGNOL, L. e ALVES, L. E. **Qualidade das plantas aromáticas comercializadas em Curitiba-PR.** In: Simpósio de Plantas Medicinais do Brasil, 15., Águas de Lindóia, 1998. Programa e Resumos. Águas de Lindóia, Conselho Nacional de Desenvolvimento Científico e Tecnológico, p.199, 1998.

DAVIDSON, P. M., Chemical preservatives and natural antimicrobial compounds, 29: 520-556. In: DOYLE, M. P.; BEUCHAT, L. R.; MONTVILLE, T. J. (eds), **Food Microbiology fundamentals and frontiers**, p.768, 1997.

DEFILIPPS, R. Conservation of Brazilian Medicinal Plants. In: **Biological Conservation Newsletter**. N.193, January 2001. Plant Conservation Unit. Department of Botany. Smithsonian National Museum of Natural History, 2001.

DEL PONTE, E.M.; GODOY, C.V.; Li. X; YANG, X.B. Predicting severity of Asian soybean rust epidemics with empirical rainfall models. **Phytopathology** 96:797-803, 2006.

DINIZ, M. F. F. M.; OLIVEIRA, R. A. G.; MALTA JUNIOR, A. **Das plantas medicinais aos fitoterápicos:** abordagem multidisciplinar. 2ª ed. João Pessoa: UFPB/CCS, p.131, 1997.

DUARTE, M.C.T. **Atividade antimicrobiana de plantas medicinais e aromáticas utilizadas no Brasil**, Universidade Estadual de Campinas Centro Pluridisciplinar de Pesquisas Químicas, Biológicas e Agrícolas DMB – Divisão de Microbiologia. Disponível em: <http:/www.multiciencia.unicamp.br/art05> Campinas, SP, 2006. Acesso em 29 abril. 2012

ELISABETSKY,E. New directions in ethnopharmacology. **Journal of Ethnobiology**, 6(1): 121-128, 1986.

ESPÍNOLA, E. B. **Uso seguro e racional de produtos fitoterápicos**. Revista Racine, p. 52-53, junho, 1997.

ESTEVES, J. A.; CABRITA, J. D.; NOBRE, G. N. **Micologia médica**. 2ª Ed. Porto (Lisboa): Fundação Calouste Gulbenkian, p.341, 1990.

ETKIN, N.L. **Etnopharmacology:** Biological and behavioral perspectives in the study of indigenous medicines. p. 149-158. In: T.M. Johnson and C.F. Sargent (eds.). **Medical antropology:** A handbook of theory and method. New York, Greenwood Press, 1990.

FARMACOPÉIA BRASILEIRA. 4ª ed. Brasília: Atheneu Editora São Paulo Ltda, p.174, 2005.

FARNSWORTH, N. R.; AKERELE, O.; BINGEL, A. S.; SOEJARTO, S. S.; GUO, Z. G. **Medicinal plants in therapy**. Bulletin of the World Health Organization, v. 63, n. 6, p. 965-981, 1985.

FERREIRA, S. H. Medicamentos a partir de plantas medicinais no Brasil. **Academia Brasileira de Ciências**, p. 133, 1998.

FIGUEREDO, C.A. **Fitoterapia**, NEPHF, p. 156, 2006.

FISCHER, D. C. H.; SAITO, T. Contaminação microbiana em fitoterápicos. **Revista de Farmácia e Bioquímica.** USP, São Paulo, v. 24, n. 2, p. 143-144, 1988.

FISCHER, D. C. H.; OHARA, M. T.; SAITO, T. Contaminação microbiana em medicamentos fitoterápicos sob a forma sólida. **Revista de Farmácia e Bioquímica**. USP, São Paulo, v.29,n.2, p.81-88, 1993.

FORD, R.I. Ethnobotany: historical diversity and synthesis. In: R.I. Ford; M. Hodge and W.L. Merril (eds.). **The nature and status of ethnobotany.** Annals of Arnold Arboretum. Michigan: Museum of Anthropology, University of Michigan, **Anthropological Papers** 67: 33-49, 1978.

GANDHI, D. N. e GHODEKAR, D. R., Antibacterial activity of garlic extract against lactic acid bacteria and contaminantes of fermented milks. **Indian J. Dairy Science**. 41: 511-512, 1988.

HAMILTON, A.C.; SHENGJI, P.; KESSY, J.; KHAN, A.A.; LAGOS-WITTE, S.; SHINWARI, Z.K. **The purposes and teaching of Applied Ethnobotany**. Godalming, People and Plants working paper. 11, 2003.

HAWKSWORTH, D.L.; KIRK, P.M.; SUTTON, B.C.; PEGLER, D.N. **Ainsworth and Bisby's dictionary of the fungi**. 8. edition Wallingford:CAB international, p. 650, 1995.

HERRICK, J.W. The symbolic roots of three potent Iroquois medicinal plants. p. 134-155. In: L. Romanucci- Ross; D.E. Moerman and L.R. Tancredi (eds.). **The antropology of medicine:** From culture to method. South Hadley, J.F. Bergin ,1983.

HITOKOTO, H.; MOROZUMI; S., WAUKE, T.; SAKAI, S.; KURATA,H. Inhibitory effects of spices on growth and toxin production of toxigenic fungi. **Applied and Environmental Microbiology**, 39: 818-822, 1980.

JAYAPRAKASHA, G.K.; RAO L.J.; SAKARIAH, K.K. **Chemical composition of volatile oil from *Cinnamomum zeylanicum*,** p.57, 2003.

JORGE, S.S.A. **Plantas Medicinais**. Coletânea de Saberes, p.85, 2009.

KALKASLIEF-SOUZA, S. B.;KIKUCHI, I. S.; MANSANO, R. D. Microbial decontamination study of some medicinal plants by Plasma treatment. **Acta Horticulturae**, Wageningen, v. 826, p. 205-211, 2009.

KARL, A.C; SOUZA, R.M e MATTOS, J.K.A. Patogenicidade de *Meloidogyne janica* em quatros espécies de plantas medicinais. **Horticultura Brasileira** p.118-121, 1997.

KNEIFEL, W. Microbial contamination of medicinal plants. A review. **Planta Medica**, v68, p.5-15, 2002.

KHAN, A; SAFDAR, M.; KHAN, M.M.A.; KHATTAK, K.N.; ANDERSON, R.A. *Cinnamom* improves glucose and lipids of people with type 2 diabetes. **Diabetes Care**, p.26, 2003.

LIMA, C.S.;SOUZA, P.E. de; BOTELHO, A.O. Fungos da família *Pucciniaceae* causadores de ferrugem em plantas medicinais. **Fitopatologia Brasileira** p.499-503. 2004.

LORENZI H.; MATOS, F.J.A. **Plantas Medicinais do Brasil: Nativas e Exóticas.** Instituto Plantarum, Nova Odessa, 2002.

LORENZI, H. **Plantas Medicinais no Brasil Nativas e Exóticas**, 2 Edição, p.544, 2008

MABROUK, S. S.; EL-SHAYEB, N. M. A.Inhibition of aflatoxin formation by some spices. Z. Lebensm. Unters. Forsch., 171: 344-347. In: GOLOB, P., MOSS, C., DALES, M., FIDGEN, A., EVANS, J.; GUDRUPS, I. (eds.). **The Use of Spices and Medicinals as Bioactive Protectants for Grains**. FAO Agric. Serv. Bull., p.137, 1980.

MACIEL,M.A.M.; PINTO,A.C.; VEIGA JÚNIOR,V.F.; GRYBERG,N.F. **Plantas medicinais:** a necessidade de estudos multidisciplinares. Universidade Federal do Rio de Janeiro, Rio de Janeiro- RJ, *Quim. Nova,* Vol. 25, No. 3, p.429-438, 2002.

MADUEÑO BOX, M. **Cultivo de plantas medicinais**. Madrid: Labor, p.490, 1973

MARTIN, G.J. **Ethnobotany - A methods manual**. London, Ed. Chapman e Hall, p.268, 1995.

MARTINS, E.R; CASTRO, D.M. de; CASTELLANI, D.C.; DIAS, J.E. **Plantas Medicinais**. Viçosa: UFV, 1995.

MARTINS, E. R.; CASTRO, D.M.; CASTELLANI, D.C.; DIAS, J.E. **Plantas Medicinais**. Viçosa/ MG: UFV, p. 220, 2003.

MATOS, F. J. A; LORENZI, H. **Plantas Medicinais no Brasil - Nativas e Exóticas.** Nova Odessa, SP: Instituto Platarum, 2002.

MATOS, F.J.A. **Farmácias vivas**. 3ª Ed., Fortaleza: EUFC, 1998.

MATOS, F. J. A. **Plantas medicinais: guia de seleção e emprego de plantas usadas em fitoterapia no nordeste do Brasil**. 2ª ed. Fortaleza: Imprensa Universitária-UFC, p.344, 2000.

MEZZARI, A.; CAUDURO, P. F. **Micologia no laboratório**. Editora Sagra, Brasil, Porto Alegre, p.139, 1996.

MELCHING, J.S; DOWLER, W.M; KOOGLE, D.L; ROYER, M.H. Effect of duration, frequency, and temperature of leaf wetness period on soybean rust. **Plant Disease** 73:117-122, 1989.

MELO, J. G.; MARTINS, J. D. G. R.; AMORIM, E. L. C. Qualidade de produtos a base de plantas medicinais comercializadas no Brazil: castanha da índia (*Aesculus hippocastanum* L.; capim limão (*Cymbopogon citratus* (D.C.) Stapf.); e centela (*Centela asiática* (L.) Urban) **Acta Botânica Brasílica**, Brasília, v. 21, n. 1, 2007.

MENDONÇA, R.; FELFILI, J.M.; WALTER, B.M.T.; SILVA JÚNIOR, M.C.; REZENDE, A.V.; FILGUEIRAS, T.S.; NOGUEIRA, P.E.N. **Flora vascular do Cerrado**, p. 287-556, 1998.

MENDONÇA, M.B.; HIDALGO, A.F.; CHAVES, F.C.M. Isolamento e identificação de fungos com potencial patogênico para a saúde humana em material vegetal de uso medicinal comercializado em Manaus. **Horticultura Brasileira**, p. 1208-1214, 2009.

MEYERS, N. Tropical moist forests; over-exploited and under-utilized? **Forest Ecology and Management** , 1983.

MING, L.C. Influência de diferentes níveis de adubação orgânica na produção de biomassa e teor de óleos essenciais de *Lippia alba* (Mill.) N.E.Br. -Verbenaceae. **Dissertação de mestrado**. Curitiba: UFPR, p.206, 1992.

MING, L.C.; SILVA, S.M.P; SILVA, M.A.S.; HIDALGO, A.H; MARCHESE, J.A.; CHAVES, F.C.M. **Departamento de Produção Vegetal Setor Horticultura.** Faculdade de Ciências Agronômicas UNESP – Botucatu – SP, 2009

MORGAN, R. **Enciclopédia das ervas e plantas medicinais.** São Paulo: Hemus, p.555, 1989.

NETO, F.R.G.; et al., Estudo Etnobotânico de plantas medicinais utilizadas pela Comunidade do Sisal no município de Catu, Bahia, Brasil. **Revista Brasileira de Plantas Medicinais**, v.16, n.4, p.856-865, 2014.

NUNES, G.P.; SILVA, M.F; RESENDE, U.M.; SIQUEIRA, J.M. de.Plantas medicinais comercializadas por raizeiros no Centro de Campo Grande Mato Grosso do Sul. **Revista Brasileira de Farmacognosia**, p 82-83, 2003

OLIVEIRA, C. A. F.; GERMANO, P.M. Aflatoxinas: conceitos sobremecanismos de toxicidade e seu envolvimento na etiologia do câncer hepático celular. **Revista de Saúde Pública**, v. 31, n. 4, p. 417-424, 1997.

PANIZZA, S. **Plantas que curam.** São Paulo: IBRASA, p.120, 1997.

PLOTKIN, M. J. **Traditional knowledge of medicinal plants:** the search for new jungle medicines. In: AKERELE, O.; HEYWOOD, V.; SYNGE, H. **Conservation of medicinal plants**. Cambridge; Cambridge University Press, p. 53-64, 1991.

PRINCIPE, P.P. **The value of biological diversity among medicinal plants.** Paris, Environment Directorate, Organization for Economic Cooperation and Development, 1985.

PRUTHI, J. S., **Spices and condiments:** chemistry, microbiology, technology. Advances in food research, suppl. 4, Academic Press, New York, p.463, 1980.

RITTER, M. R.; SOBIERAJSKY, G. R.; SCHENKEL, E. P.; MENTZ, L. A. C. Plantas usadas como medicinais no município de Ipê. **Revista Brasileira de Farmacognosia**, RS, Brasil. vol.12, n. 2, p. 51-62, 2002.

ROITMAN, I.; TRAVASSOS, L. R.; AZEVEDO, J. L. **Tratado de microbiologia**. São Paulo: Manole, p.186, 1988.

RUSSOMANNO, O.M.R.; KRUPPA, P.C. Doenças fúngicas das plantas medicinais, aromáticas e condimentares. **Divulgação Técnica Biológica**, Instituto Biológico, Centro de Pesquisa e Desenvolvimento de Sanidade Vegetal, São Paulo (SP), p. 31-37, junho, 2010.

SAAD, B; AZAIZEH, H; SAID, O. **Tradition and Perspectives of Arab Herbal Medicine:** A Review, 2(4): 475-479, october, 2005.

SANGUINETTI, E. E. **Plantas que curam.** 2 ed. Editora Rígel, Porto Alegre p. 208, 1989.

SANTOS, P. R. V.; OLIVEIRA, A. C. X.; TOMASSINI, T. C. B. Controle microbiológico de produtos fitoterápicos. **Revista Farmacêutica Bioquímica.** Universidade São Paulo, v. 31,n.1, p.35-38, 1995.

SATOMI, L. C.; SORIANI, R. R.; PINTO, T. J. A. Descontaminação de drogas vegetais empregando radiação gamma e óxido de etileno: aspectos microbianos e químicos. **Revista Brasileira de Ciência Farmacêuticas**, São Paulo, v. 41, n. 4, p. 445-450, 2005.

SCHEFFER, M.C. Roteiro para estudo de aspectos agronômicos das plantas medicinais selecionadas pela fitoterapia do SUS-PR/ CEMEPAR. **SOB Informa** v. 10-11, n. 1-2, p. 29- 31, 1992.

SCHULTES, R.E. ;REIS, S.V. (eds.). **Ethnobotny: evolution of a discipline**. Cambridge, Timber Press, 1995.

SHAN, B; CAI, Y.Z.; SUN, M.; CORKE, H. Antioxidant Capacity of 26 spices extracts and characterization of their phenolic constituents. **J Agriculture Food Chemist.**, p.53, 2005.

SILVA, I.; FRANCO, S.L.; MOLINARI, S.L. **Noções sobre o organismo humano e utilização de plantas medicinais**. 3ª Ed., Cascavel (PR), 1995.

SILVA,C.G.; et al., Levantamento etnobotânico de plantas medicinais em área de Caatinga na comunidade do Sítio Nazaré, município de Milagres, Ceará, Brasil. **Revista Brasileira de Plantas Medicinais**, v.17, n.1, p.133-142, 2015.

SOUZA, M.P.; MATOS, N.E.O.; MATOS, F.J.A. **Constituintes Químicos de Plantas Medicinais Brasileiras**. Imprensa Universitária/UFC, Fortaleza, 1991.

SOUZA, R.M.; MATTOS, J.K.A; KARL, A.C. Avaliação preliminar da reação de plantas medicinais a *Meloidogyne javanica* e M. *incognita*. **Horticultura Brasileira** vol.13: p. 209-211,1995.

SUZUKI, O. Mercado de medicamentos fitoterápicos no Brasil. In: SCHULZ, V.; HÄNSEL, R.; TYLER, V. E. **Revista Brasileira de Plantas Medicinais** 4. ed. São Paulo: Manole, p. 94- 99, 2002.

TORTORA, G. J.; FUNKE, B. R.; CASE, C. L. **Microbiologia**. 6. ed. Porto Alegre: Artmed, 2000.

TSCHANZ, A.T. Soybean rust epidemiology: Final Report. **Asian Vegetable Research and Development Center**, Shanhua, Taiwan, 1984.

URBEN, A.F.;MATTOS, J.K.A.; MENDES, M.A.S. Fungos associados a manchas de folhas em plantas de uso medicinal. **Fitopatologia Brasileira** Distrito Federal, vol12: p. 390-394. 1987.

VEIGA JÚNIOR, F.V.; PINTO, A.C; MACIEL, M.A. **Plantas medicinais:** cura segura? Universidade Federal do Rio Grande do Norte, Campus Universitário, Natal – RN, vol. 28, no. 3, p. 519-528, 2005

VIEIRA, L.S; SOUSA, R.S; LEMOS, J.R. Plantas medicinais conhecidas por especialistas locais de uma comunidade rural maranhense. **Revista Brasileira de Plantas Medicinais**. Campinas, v.17, n.4, supl. III, p.1061-1068, 2015.

ZARONI, M.; PONTAROLO, R.; ABRÃO, W. S. M. Qualidade microbiológica das plantas medicinais produzidas no Estado do Paraná. **Revista Brasileira de Farmacognosia**, São Paulo, v. 14, n. 1, 2004.

Printed by Books on Demand GmbH, Norderstedt / Germany